YOUR KNOWLEDGE HAS VALUE

- We will publish your bachelor's and master's thesis, essays and papers

- Your own eBook and book - sold worldwide in all relevant shops

- Earn money with each sale

Upload your text at www.GRIN.com and publish for free

Manoj Parakhia, B.J. Malviya, R.S. Tomar, R.M. Dhingani, B.A. Golakiya

Economical and effortless Medium for the isolation and growth of fungi.

GRIN Verlag

Bibliografische Information der Deutschen Nationalbibliothek:

Die Deutsche Bibliothek verzeichnet diese Publikation in der Deutschen Nationalbibliografie; detaillierte bibliografische Daten sind im Internet über http://dnb.d-nb.de/ abrufbar.

Dieses Werk sowie alle darin enthaltenen einzelnen Beiträge und Abbildungen sind urheberrechtlich geschützt. Jede Verwertung, die nicht ausdrücklich vom Urheberrechtsschutz zugelassen ist, bedarf der vorherigen Zustimmung des Verlages. Das gilt insbesondere für Vervielfältigungen, Bearbeitungen, Übersetzungen, Mikroverfilmungen, Auswertungen durch Datenbanken und für die Einspeicherung und Verarbeitung in elektronische Systeme. Alle Rechte, auch die des auszugsweisen Nachdrucks, der fotomechanischen Wiedergabe (einschließlich Mikrokopie) sowie der Auswertung durch Datenbanken oder ähnliche Einrichtungen, vorbehalten.

Imprint:

Copyright © 2013 GRIN Verlag GmbH
Druck und Bindung: Books on Demand GmbH, Norderstedt Germany
ISBN: 978-3-656-47500-2

This book at GRIN:

http://www.grin.com/en/e-book/231059/economical-and-effortless-medium-for-the-isolation-and-growth-of-fungi

GRIN - Your knowledge has value

Der GRIN Verlag publiziert seit 1998 wissenschaftliche Arbeiten von Studenten, Hochschullehrern und anderen Akademikern als eBook und gedrucktes Buch. Die Verlagswebsite www.grin.com ist die ideale Plattform zur Veröffentlichung von Hausarbeiten, Abschlussarbeiten, wissenschaftlichen Aufsätzen, Dissertationen und Fachbüchern.

Visit us on the internet:

http://www.grin.com/

http://www.facebook.com/grincom

http://www.twitter.com/grin_com

Economical and effortless Medium for the isolation and growth of fungi.

*M. V. Parakhia, B. J. Malviya, R. S. Tomar, R. M. Dhingani and B. A. Golakiya
Department of Biotechnology, Junagadh Agricultural University, Junagadh – 362 001, Gujarat, India.

Abstract

For a long time used many growth media like Potato Dextroxe agar, Capazedx agar, Rose Bengal agar for the isolation and identification of various type of fungi like *Aspergillus, Penicillium, Tricoderma* etc. but the preparation of above said medium procedure is very lengthy and the components are used in this medium is so coastally except Potato Dextroxe agar (PDA). the PDA medium widely used for the isolation of fungi because it is easy to prepare and cheap ingredients. But another medium Tomato agar we prepared is more easy to prepare than PDA and more Cheap than the PDA. The growth pattern of fungi on Tomato Agar compare to Growth pattern of fungi on PDA was not significantly difference so the tomato agar medium will be used for isolation of fungi alternate PDA which is cheap medium for the isolation of fungi and may widely used in research laboratories.

Key Words : Potato Dextroxe agar, Fungi, Tomato agar

***Corresponding Author**
Manoj V. Parakhia
Assistant Professor

Introduction

The word "**tomato**" may refer to the plant (*Solanum lycopersicum*) or the edible, typically red, fruit that it bears. The tomato fruit is consumed in diverse ways, including raw, as an ingredient in many dishes and sauces, and in drinks. While it is botanically a fruit, it is considered a vegetable for culinary purposes, which has caused some confusion. The vegetable is rich in lycopene, which may have beneficial health effects.

Red tomatoes, raw Nutritional value per 100 g	
Energy	74 kJ (18 kcal)
Carbohydrates	3.9 g
Sugars	2.6 g
Dietary fiber	1.2 g
Fat	0.2 g
Protein	0.9 g
Water	94.5 g
Vitamin A equiv	42 μg (5%)
lutein and zeaxanthin	123 μg
Vitamin C	14 mg (17%)
Vitamin E	0.54 mg (4%)
Potassium	237 mg (5%)
Source: USDA Nutrient Database	

The cost of Potato Dextrose Agar was Rs. 368/ 100gm and Rs. 1490/ 500gm (Hi-media catalog year- 2010-11) and it is recommended that 39 gm used for preparation of one liter of

PDA. So the cost for preparation of 100ml of PDA was Rs. 120-144. Manually preparation of PDA in laboratory cost was 80-90 times less than the used prepared one and the procedure for preparation of PDA was complex and time consuming. For the alternate of this the Tomato agar medium will be used in place of PDA because preparation procedure was very easy and cheap method.

Raw tomato which are not suitable for the human consumption like some tomato ripening high, break of tomato during picking those types of tomato throws in waste which tomato was used for the preparation of tomato agar so this will help to spreading of pollution and benefit to

Tomato juice agar is a well known medium used in Mycology laboratories, simple to prepare, cheap and that now, it can have a double usage indicating the sexual differentiation of yeasts and as a presumptive medium for differentiation between *C. albicans* and *C. dubliniensis*[1]

Tomato juice agar or V-8 juice agar is a well known medium widely used for ascospore formation in yeasts as *Saccharomyces cerevisae* and *Hansenula anomala* [2, 3, 4]

Second important feature of Tomato agar is that no need to add antibacterial antibiotics which we can used during the preparation of potato dextrose agar for the inhibit the growth of bacteria while tomato agar have acidic pH so no needed. this will also reduce the cost of medium.

Material and Methods
Culture: Seven different strains of *Trichoderma* collected for NBAIM, Kasuar, Banglore. Four Phytopathogenic fungi obtained from Department of plant Pathology, JAU, Junagadh.(Gujarat).
Media: Potato Dextroxe agar, Tomato agar
Methods: The preparation of tomato agar is very easy

As per the laboratory procedure the preparation of potato Dextrose agar is that first boiled the potato than extract of potato should be used for the preparation of Potato Dextrose agar. Following proportional in 100ml PDA

Potato Dextrose Agar	**Tomato Agar**
Potato infusion................30gm	Tomato Pulp 30gm
Dextrose........................2gm	Distilled water................100ml
Distilled water...............100ml	Agar powder...................2gm
Agar powder...................2gm	p^H .... 5.0 to 6.0
p^H .... 6.5 to 6.8	

The preparation of tomato agar was very essay as compare to any other methods for the preparation growth medium for isolation and identification of bacteria.

Procedure:
1. Weight a 30 gm of tomato pieces . then crush in 100ml Distilled Water
2. Filter the tomato pulp using whatman filter paper for the remove of seeds and other derbies.
3. Take a filtered material in measuring cylinder and make final volume 100ml with D/W
4. Add two gram of Agar-Agar powder in 100ml tomato pulp.
5. Check the pH of Medium
6. Plug the flash and autoclave it at 15lbs pressure and 121^0C for 15 minutes
7. After autoclaving pour on sterile Petri dishes in sterile condition.
8. Solidified it.
9. Tomato agar plates ready to use for isolation of fungi.

Result and Discussion

Small piece of fungal growth put on PDA and Tomato agar plate. Incubate plates at room temperature (28-30 0C) for 72 hrs and 96 hrs. The growth pattern of different species of fungi on Tomato agar plate and Potato dextrose agar are as below:

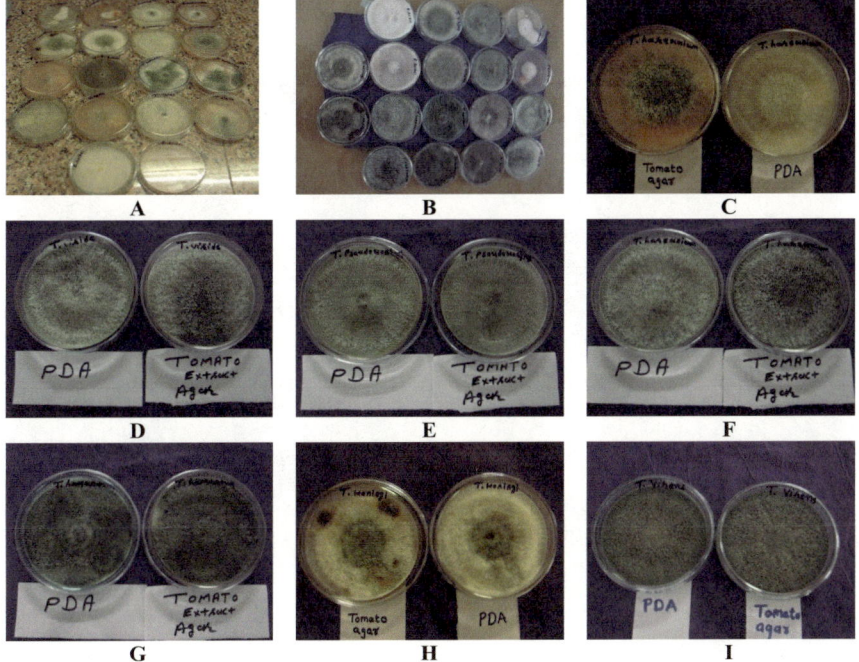

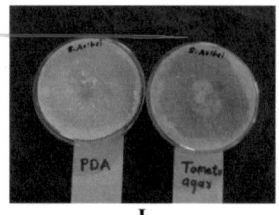 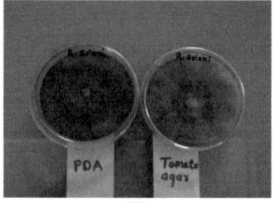 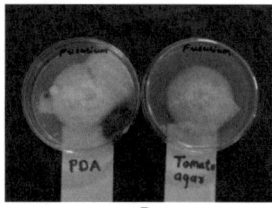

          J   K   L

A   Growth of Diffeent strain and species of fungi on Tomato agar and PDA after 48 hrs incubation at room temperature

B   Growth of Diffeent strain and species of fungi on Tomato agar and PDA after 96 hrs incubation at room temperature

C   Growth of *Trichoderma harzinium* on Tomato agar and PDA after 72 hrs incubation at room temperature

D   Growth of *Trichoderma virens* on Tomato agar and PDA after 96 hrs incubation at room temperature

E   Growth of *Trichoderma pseudokonigii.* on Tomato agar and PDA after 96 hrs incubation at room temperature

F   Growth of *Trichoderma harzinium* on Tomato agar and PDA after 96 hrs incubation at room temperature

G   Growth of *Trichoderma Hamantum* on Tomato agar and PDA after 96 hrs incubation at room temperature

H   Growth of *Trichoderma Koningi* on Tomato agar and PDA after 96 hrs incubation at room temperature

I   Growth of *Trichoderma virens* on Tomato agar and PDA after 96 hrs incubation at room temperature

J   Growth of *Scelosium rolfsii* on Tomato agar and PDA after 72 hrs incubation at room temperature

K   Growth of *Rhizoctina solani* on Tomato agar and PDA after 72 hrs incubation at room temperature

L   Growth of *Fusariun oxysphorium* on Tomato agar and PDA after 96 hrs incubation at room temperature

      Above result shows that no significant different found in the growth pattern of different types of fungi on PDA and TA.

Conclusion

From above result conclude that the Tomato juice agar was easy to prepare and cheap medium compare to other medium available in market for the isolation and cultivation of fungi. tomato agar also used for the isolation of *Lactobacillus Spp*. of bacteria.

References

1. SYDNEY HARTZ ALVES, ÉRICO SILVA DE LORETO & CARLOS EDUARDO LINARES, 2006. Comparison Among Tomato Juice Agar With Other Three Media For Differentiation of Candida Dubliniensis From Candida AlbicansRev. Inst. Med. trop. S. Paulo 48(3):119-121.
2. ALEXOPOULOS, C.J.; MIMS, C.W. & BLACKWELL, M., 1996. Introductory Mycology. 4. ed. New York, John Wiley.
3. JENNINGS, D.H. & LYSEK, G. 1999. Fungal Biology: understanding the fungal lifestyle. 2. ed. New York, Springer-Verlag.
4. LARONE, D.H.,1995. Medically important fungi: a guide to identification. 3.ed. Washington, ASM Press.